MAMMALS

Animal Group Science Book For Kids
Children's Zoology Books Edition

Speedy Publishing LLC
40 E. Main St. #1156
Newark, DE 19711
www.speedypublishing.com

Mammals are a particular class of animal. All mammals produce milk for their babies to drink.

The lion is the second largest cat in the world. Lions are found in savannas, grasslands, dense bush and woodlands.

Lions are very social cats and live in groups called prides. A lion can run for short distances at 50 mph and leap as far as 36 feet.

The koala is an arboreal herbivorous marsupial native to Australia. Koalas typically inhabit open eucalypt woodlands.

Koalas have large noses that are coloured pink or black. A koala mother usually gives birth to one joey at a time.

The giraffe is the tallest living terrestrial animal. Giraffe's live in African savannas, grasslands or open woodlands.

Giraffes have bluish-purple tongues which are tough and covered in bristly hair to help them with eating the thorny trees.

Elephants are the largest land-living mammal in the world. It is characterized by its highly dexterous trunk, long curved tusks, and massive ears.

Elephants use their trunk for smelling, breathing, detecting vibrations, caressing their young, sucking up water, and grasping objects.

The polar bear are the world's largest land predators. Male polar bears can weigh up to 680 kg.

Polar bears live along shores and on sea ice in the icy cold Arctic. Seals make up most of a polar bears diet.

The hippopotamus live in East Africa South of the Sahara. They are considered a dangerous animal in Africa.

Hippopotamus spend a large amount of time in water such as rivers, lakes and swamps.

Dolphins are a diverse group of fully aquatic marine mammals. Dolphins live in schools or pods of up to 12 individuals.

Dolphins use a blowhole on top of their heads to breathe. Dolphins give birth to babies underwater.

The orangutans are currently found in the rainforests of Borneo and Sumatra. Orangutans spend most of their time in trees.

Orangutans are among the most intelligent primates they use a variety of sophisticated tools.

Bats are the only mammals naturally capable of true and sustained flight. There are over 1000 different bat species.

Bats are nocturnal. Most bats feed on insects, while others eat fruit.